APPLES

J.F. QUICK

look!

APPLES

CREATIVE EDUCATION • CREATIVE PAPERBACKS

Published by Creative Education
and Creative Paperbacks
P.O. Box 227, Mankato, Minnesota 56002
Creative Education and Creative Paperbacks are imprints of
The Creative Company
www.thecreativecompany.us

Design and art direction by Blue Design
Edited by Joe Tischler

Photographs by Dreamstime (Alexandr Vasilyev, Ivandzyuba, Ivandzyuba, Justin Skinner, Photozirka), Getty (BehindTheLens, benedek, Green_Leaf, ICHIRO KATAKAI/amanaimagesRF, jenifoto, Kateryna, Bibro, Olha_stock, pixhook, romrodinka, Wataru Yanagida, Wataru Yanagida, Ольга Симонова), Shutterstock (de2marco, Fotofermer, Irena Misevic, Maks Narodenko, Olga Popova, Picture Partners, Ralf Kleemann, wabeno), SuperStock (Animals Animals, Laurence Mouton/PhotoAlto), Unsplash (Kari Shea)

Copyright © 2025 Creative Education, Creative Paperbacks
International copyright reserved in all countries.
No part of this book may be reproduced in any form without written permission from the publisher.

Library of Congress Cataloging-in-Publication Data

Names: Quick, J. F., author.
Title: Apples / by J.F. Quick.
Description: Mankato, Minnesota : Creative Education and Creative Paperbacks, [2025] | Series: Look! | Includes bibliographical references and index. | Audience: Ages 6-9 | Audience: Grades 2-3 | Summary: "An engaging story of the life cycle of apples, this middle-grade book features eye-catching photographs, insightfully simple language, and biological facts about this tasty fruit"-- Provided by publisher.
Identifiers: LCCN 2023047344 (print) | LCCN 2023047345 (ebook) | ISBN 9781640268500 (library binding) | ISBN 9781682774007 (paperback) | ISBN 9798889890362 (ebook)
Subjects: LCSH: Apples--Juvenile literature. | Apples--Life cycles--Juvenile literature.
Classification: LCC SB363 .Q53 2025 (print) | LCC SB363 (ebook) | DDC 634/.11--dc23/eng/20231211
LC record available at https://lccn.loc.gov/2023047344
LC ebook record available at https://lccn.loc.gov/2023047345

Printed in China

TABLE OF CONTENTS

Apples 8
Starting From Seeds 11
Branching Out 12
Beautiful Blossoms 14
Help From Bees 15
Ripe and Ready 16
Resting for Next Year 19
The Cycle Continues 20
Apple Tree Cycles 21
Johnny Appleseed 23
Websites 24
Read More 24
Index 24

Apples are fruits that grow on trees. They make a tasty, healthy snack. They can be made into sauces, pies, juices, and cider, too.

Apple trees grow on every continent but Antarctica. They can grow in forests, parks, and in people's back yards. Most apples grow in places called orchards. In orchards, many apple trees are planted close together.

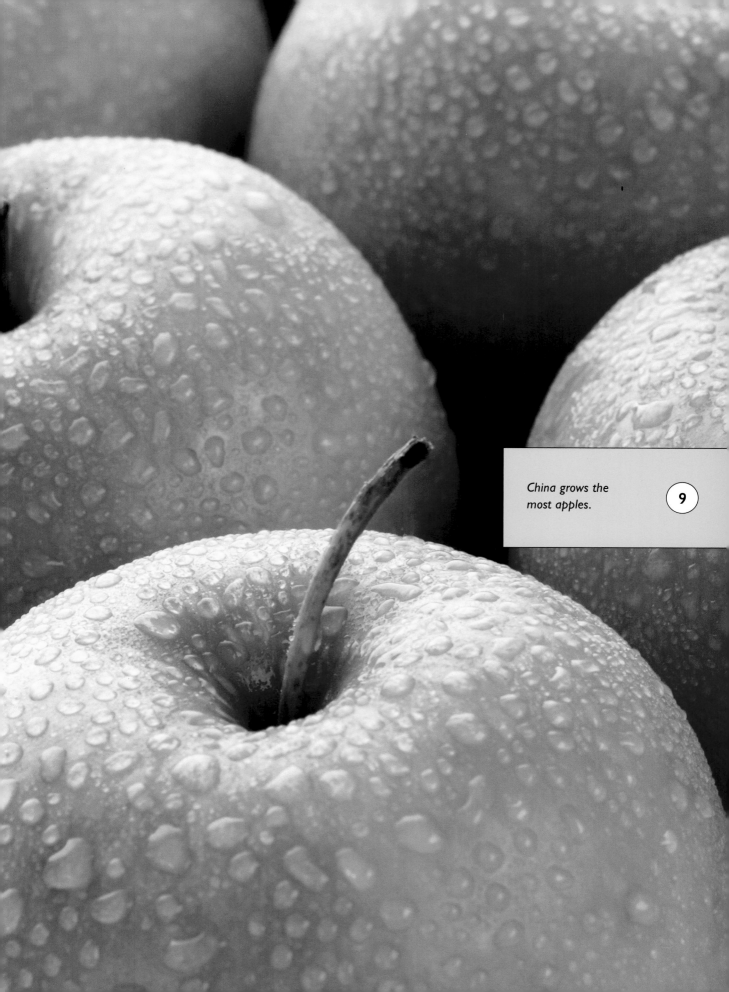

China grows the most apples.

9

10 Every apple has five sections for seeds. An apple usually contains between five and eight seeds.

STARTING FROM SEEDS

An apple tree starts with an apple seed in the soil. The seed may be planted or dropped.

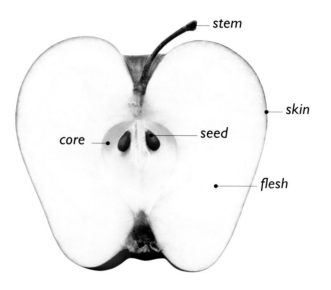

When spring rain soaks the seed, it starts to grow. **Roots** grow down into the earth. A stem grows up out of the ground. On the stem, leaves grow. The young tree is called a seedling.

roots: the parts of a plant that grow underground

BRANCHING OUT

The seedling's stem grows wider and stronger. It becomes the tree's trunk. Branches start to grow from the trunk.

Leaves grow on the branches. The leaves use light, water, and air to make food for the tree. The seedling grows into a strong young tree. The young tree is called a sapling.

The sapling's tough bark protects its soft insides from germs, animals, and the cold.

BEAUTIFUL BLOSSOMS

The sapling keeps growing for 5 to 10 years. One spring, when the tree is ready, tiny bumps called buds will grow all over the tree's branches.

The buds open into pinkish white blossoms. The middle of each blossom is covered with sticky yellow **pollen**.

pollen: tiny yellow grains produced by flowers that help plants make new plants

HELP FROM BEES

Bees land on apple blossoms. They drink a sweet liquid called nectar. Tiny bits of pollen stick to their fuzzy bodies. The bees carry pollen to other trees.

pollinate: to carry pollen from one flower to another

Apple trees need pollen from other apple trees to **pollinate** their blossoms. From a pollinated blossom, an apple starts to grow. The apples grow all summer.

RIPE AND READY

Apples are ready to eat when you can easily pull them off a branch. This happens in autumn. This is when orchard workers will **harvest** their apples. Many people visit orchards in autumn to pick apples. Picking apples is a fun way to spend an autumn day.

MANY APPLE TYPES TO TRY

GALA
Color: *Red*
Flavor: *Mild and crisp*

GOLDEN DELICIOUS
Color: *Yellow*
Flavor: *Honey*

GRANNY SMITH
Color: *Green*
Flavor: *Tart*

FUJI
Color: *Red*
Flavor: *Sweet*

harvest: to gather fully grown crops

People can use apple pickers to reach fruit in the highest branches.

17

Being dormant protects the tree from winter cold.

RESTING FOR NEXT YEAR

In winter, an apple tree becomes **dormant**. Its leaves fall off. It does not grow any new parts. This helps the tree save energy. New buds inside the branches are getting ready to grow when spring arrives.

dormant: still alive but not actively growing

THE CYCLE CONTINUES

20 Every year, apple trees go through a **cycle**. They grow leaves and blossoms in spring. Apples grow all summer and are ripe in autumn. The trees lose their leaves for winter and become dormant.

The cycle repeats many times. An apple tree can live more than 100 years before it dies.

cycle: a pattern that happens over and over in the same order

APPLE TREE CYCLES

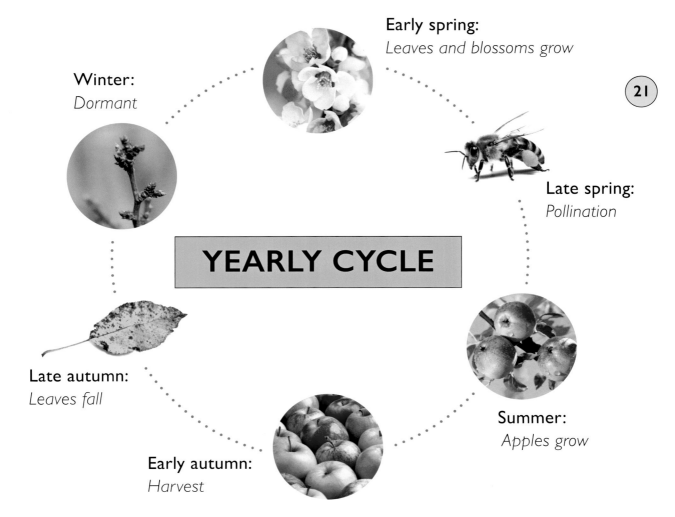

YEARLY CYCLE

Early spring: *Leaves and blossoms grow*

Late spring: *Pollination*

Summer: *Apples grow*

Early autumn: *Harvest*

Late autumn: *Leaves fall*

Winter: *Dormant*

APPLE TREE CYCLES

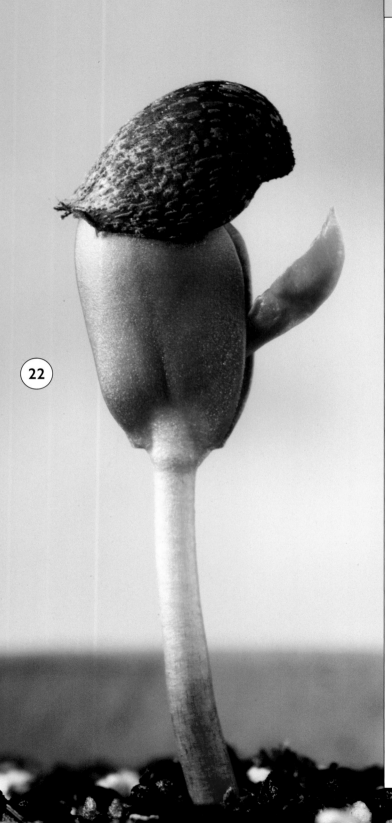

LIFE CYCLE

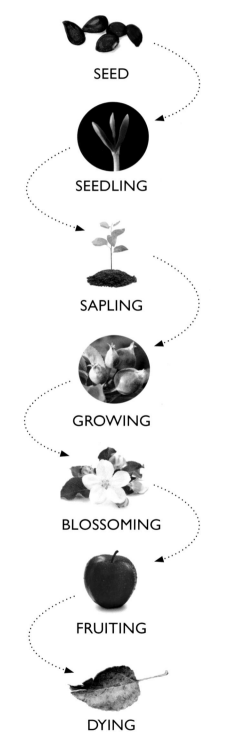

SEED

SEEDLING

SAPLING

GROWING

BLOSSOMING

FRUITING

DYING

JOHNNY APPLESEED

In the 1800s, a man named John Chapman walked through many places in the United States. He carried apple seeds in a bag. He planted them in the soil. He wanted to make sure there were apple trees everywhere. With hard work, he made the land bloom with tasty apples. People called him "Johnny Appleseed."

Johnny Appleseed was known to walk across the country without shoes.

WEBSITES

Crispy's Apple Stand
https://waapple.org/did-you-know/
Check out apple facts, activities, and recipes.

Enchanted Learning: Johnny Appleseed
https://www.enchantedlearning.com/school/USA/people/Apple-seedindex.shtml
Learn more about Johnny Appleseed.

READ MORE

Charles, Beth. *How to Grow an Apple Pie.* Morton Grove, I.L.: Albert Whitman & Company, 2020.

Knowlton, Laurie Nazzaro. *Who Knew? Under the Apple Tree.* Mankato, Minnesota: Amicus Ink, 2021.

INDEX

autumn 16, 20, 21
bees 15
blossoms 14, 20–23
branches 12, 14, 17, 19
buds 14, 19
cycles 20–22
dormant 18–21
foods from 8
harvesting 16, 21
leaves 11, 12, 19–21
pollination 14, 15
orchards 8, 16
parts 11, 19
range 8
saplings 12–14, 22
seeds 10, 11, 22, 23
seedlings 11, 12, 22
spring 11, 14, 19–21
summer 15, 20, 21
trunks 12
types 16
winter 18-21